Is It Alive?

Table of Contents

Marcia S. Freeman

Some things on Earth are alive.
All the other things on Earth are not.

How can you tell what is alive?

Brown Booby

All living things grow and change. You are growing right now, but you grow so slowly that you hardly notice.

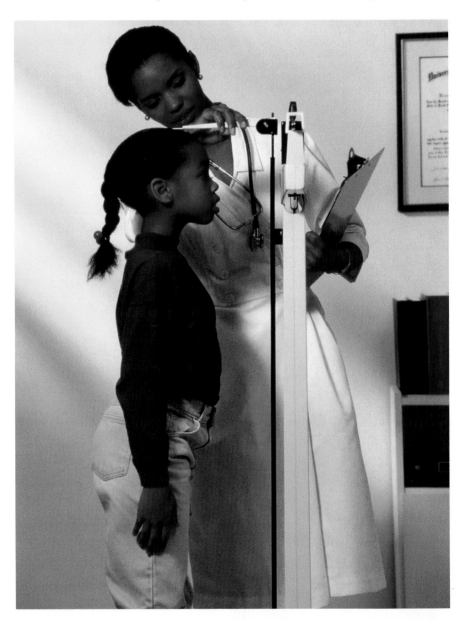

Some things, like these bulbs, grow quickly. You can see the change in a few days.

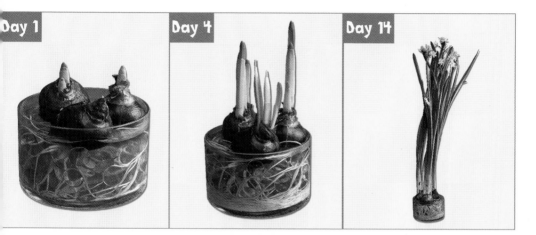

These rocks do not grow.
They are not alive.
Nonliving things do not grow.

Living things **reproduce**.
Plants make seeds or grow shoots.
Animals lay eggs or have live babies.

Butterfly laying eggs

Which of these things
are alive? Which are not?

How do you know?

All living things need food to **survive**.

Plants make their own food from sunlight, air, and water.

Animals eat plants
or other animals.

Living things need water.

Most animals drink fresh water every day.

Most plants **absorb** water through their roots.

And all living things **excrete** wastes. Wastes are left over when living things use food and water.

Only living things can move on their own. Nonliving things need something else to move them.

Which of these things are moving on their own?

What is making the other things move?

Some nonliving things were never alive.

These bottles are made of glass, which was never alive.

Other nonliving things were once alive. This cabin is made of logs. The logs came from living trees.

These leaves were alive before they fell from a tree.

All living things are made up of living parts called **cells**.

Cells are like tiny building blocks.

They are so small that you need a **microscope** to see them.

Human tongue cells

Lilac leaf cells

The best way to tell if something is alive is to see if it has cells.

But if something grows, reproduces, uses food and water, and moves on its own—you can be sure it is alive.

Is it alive?

How can you tell?

Glossary

absorb (uhb-ZOHRB): take in; soak up

cells (SELZ): the tiny units that make up all living things and things that once lived

excrete (ik-SKREET): get rid of waste from the body

microscope (MYE-kruh-skohp): an instrument with lenses that makes small things look large and helps people see tiny things such as cells

nonliving (NON-LI-ving): not alive

reproduce (ree-pruh-DOOS): make another; produce seeds, eggs, or babies

survive (sur-VYV): stay alive

Index